O. G. Pritchard

The Manufacture of Electric Light Carbons

O. G. Pritchard

The Manufacture of Electric Light Carbons

ISBN/EAN: 9783337060930

Printed in Europe, USA, Canada, Australia, Japan

Cover: Foto ©berggeist007 / pixelio.de

More available books at **www.hansebooks.com**

OF

ELECTRIC LIGHT CARBONS.

BY

O. G. PRITCHARD.

Reprinted from " THE ELECTRICIAN."

WITH ADDITIONAL CHAPTERS.

PRICE ONE SHILLING AND SIXPENCE,

Post Free 1s. 9d.

LONDON:

"THE ELECTRICIAN" PRINTING AND PUBLISHING COMPANY, LIMITED,

SALISBURY COURT, FLEET STREET, E.C.

THE MANUFACTURE

OF

ELECTRIC LIGHT

CARBONS.

BY

O. G. PRITCHARD.

Reprinted from " THE ELECTRICIAN,"
WITH ADDITIONAL CHAPTERS.

PRICE ONE SHILLING AND SIXPENCE,
Post Free 1s. 9d.

LONDON:
"THE ELECTRICIAN" PRINTING AND PUBLISHING COMPANY, LIMITED,
SALISBURY COURT, FLEET STREET, E.C.

1890.

LONDON :
PRINTED BY GEORGE TUCKER, 1, 2 AND 3, SALISBURY COURT,
FLEET STREET, E.C.

CONTENTS.

THE MANUFACTURE

OF

ELECTRIC LIGHT CARBONS.

By O. G. PRITCHARD.

INTRODUCTION.

THAT hitherto we have been unable to cope with France, Austria, Germany, or Bohemia in the production of electric-arc-carbon candles is seen by the fact that our supplies continue to be drawn from Continental manufacturers. Paris claims good makers, such as Carré, Sautter-Lemonnier, Lacombe, Gaudoin, and others, but their prices being high in great measure precludes their adoption on a large scale. Germany boasts Herren Siemens and Schmeltzer, Austria Herr Hardtmuth, and the Apostle carbon comes to us from Bohemia. These carbons have of late years been much reduced in price, and leave little or nothing to be desired. From America we obtain the Brush carbon, solid and coppered, which eventually we may expect to see superseded by the cored uncoppered carbon. Other large makers might also be mentioned, to whose specialities I need not now refer. Up to the present we stand committed to Siemens, Hardtmuth, and Schmeltzer, who, between them, and the Apostle carbons, divide the honours and the profits. There are other first-rate makers, but they are but little known to me, and have not as yet introduced their carbons on so large a scale as the German makers. Strenuous efforts have been made during the past ten years to introduce a home-made article, efficient and good, but failure alone appears to have waited upon endeavour. This has led to the idea that carbon-candle making can never become an English industry. The writer has, however, produced carbons on a scale of some magnitude which were testified by experts to be quite equal to those of the foreign makers; and he now publishes the whole process employed by him, and the results of eight years' experiments and experience, in the hope that the arc-light carbon industry may eventually be profitably introduced.

The difficulties in the way of producing a perfect carbon, like all other difficulties, vanish with close application to the subject. The various processes, so easily detailed and explained, however, have only been perfected after many years' devotion to the subject and large expenditure of funds. How not to do it is the antecedent to all discoveries; and every inventor, or would-be discoverer of hidden processes, has to learn that only by the continuous registration of negative results can he hope to achieve positive success. I think it will be instructive and also interesting to explain the many faults to which carbons are prone, to diagnose the maladies, and to point out the cause of failure, the necessary preliminary to a cure. The *ultima thule* of carbon perfection, perhaps, has not yet been achieved. Yet I am of opinion that, with the elimination of the faults hereinafter described, a carbon will be produced as near perfection as the severest critic can wish for.

I think we get the carbon and the light in their worst form in the Jablochkoff system. The varying displays of blue, red and yellow and green lights certainly afford a great choice of colouring, but are undesirable, to say the least.

This system, however, being a thing of the past, need not occupy our time. Hissing, also, may now be said to be like the dodo—dead. We come, however, to a serious defect in the little gaseous flame that is sometimes seen to travel round the point of the positive carbon whenever the arc lengthens beyond the normal, and which throws a dark shadow upon the bowl. This defect may be also enhanced by a faulty lamp not answering quickly to the change in the amount of current, and not allowing the arc to become shortened. This has been attributed to the lamp; but seeing that by removing the carbon, and placing a good carbon in circuit, the effect ceases, the fault must be attributed to the carbon, and upon investigation it will be found to arise from the presence of occluded gases in the body of the carbon. This, again, arises from the improper nature of the material employed, which renders the carbon incapable of receiving or retaining the compression to which it had to be submitted in the press ; or it may arise from want of due saturation in an after process, whereby the want of compression is got over, and the pores containing these occluded gases are filled with carbonised matter. All or some of these may be assigned as reasons.

But the original fault lies in the material employed. I make this statement with boldness, and further assert that the purest carbon is incapable of yielding a good carbon candle unless it possesses qualities which have to be imparted to it mechani-cally. The nearest approach which Nature affords for a proper material is gas-retort carbon, inasmuch as the grain is hard, good, and so well fitted to the purpose. But this material is too impure for use, and is expensive to purify, which prevents its adoption. When the preparation of material is touched upon, this will be more clearly understood. Now, the grain and character of the material, outside and beyond its purity, are matters of moment. By way of more fully illustrating my meaning, let me take the purest lampblack, perfectly suitable *per se* for carbon-making as regards purity, yet, owing to the absence of grain, it cannot be made to retain its compression ; it occludes gases, and the perfection of the carbon is destroyed, and its life considerably shortened. As far as my experience reaches, I should say that a 10-millimetre carbon (cored positive), 42 .volts, 8-ampere current, $\frac{3}{16}$in. arc, gives the highest results. If the arc lengthen to $\frac{1}{4}$in., red flames are produced in the best carbons I have tested.

Conductivity is of the greatest importance : the greater the facility with which the current passes, the better the light and the less the E.M.F. required. Thus I think I may state, without experimental data to go upon, but by observation only, that a candle of high conductivity will give as good a light with a 6-ampere current and 38 volts as one of lower conductivity using 42 volts and 8 amperes. Indeed, when the conductivity is very low, the current will not produce the arc, and the carbon becomes intensely heated. This conductivity depends entirely (given pure carbon) on the length of time to which it has been submitted to the firing process. Graphite appears to owe its peculiar condition of high conductivity to its lengthened subjection to Nature's forces, unapproachable by human efforts. Great pressure, continuous and intense heat protracted over long periods, even if possible, would prove commercially impracticable ; yet I entertain no doubt that carbon submitted to a white heat, such as would melt wrought iron, for a whole week, would produce a candle more graphitic in character and possessing higher conductivity than one that had been subject to such treatment for a less period. As I deal with the subject of firing, and my special economic arrangements for so doing, this subject will be more fully dealt with.

I would like to mention (as it may not be thought sufficient to take the raw graphite ignorant of the amount of pure carbon it contains) a method of testing this raw material. This, indeed, should be done repeatedly, and samples kept, marked with the results of the analysis. By this means experience will be gained, and the look will almost be sufficient guide. Take and pulverise a small sample taken from the bulk, and dry it well at a temperature of 380° F., so as to make sure that all the moisture has been driven off. Take one gramme of this powder and 20 grammes of oxide of lead, also well dried; mix them thoroughly together and pour them into a hard glass test-tube, five or six inches long and half-an-inch diameter. Weigh the tube and its contents carefully, and submit the tube to a white heat under a Fletcher's blowpipe until all gases are driven off and the contents completely fused. Allow the tube to cool, and weigh the residue in the tube. The weight lost is carbonic acid, the oxygen of which has been taken from the lead oxide, while the carbon is all that there was in the graphite. For every 20 parts of loss there must have been 12 parts of carbon.

Again, the duration of the positive carbon is a matter of moment, since it is *de rigueur* that the waste should not exceed 1in. per hour; and as this is now an accomplished fact, it has become a necessity. For some inscrutable reason, foreign makers supply only 10in. positives unless otherwise ordered, so that with 1in. in the holder, and something to spare to save the lamp from heating, little is left besides the 8in. required for eight hours' light. I have met with no material in its native state capable of reducing the consumption to this standard; the prepared material employed by the writer satisfies these conditions.

Another defect, and one which makes the otherwise perfect carbon valueless, is curvature. This takes place in drying and firing. If not properly or sufficiently dried, carbons will not only bend during the firing process, but break into several pieces. The method of avoiding this will find its place further on. With lampblack carbons, fully 50 per cent. will be curved, but with the prepared material described at length,

c

and with proper care taken in the drying, no anxiety need arise. Another matter of grave importance, the consideration of which, if neglected, will lead to most serious results, is the following. A carbon may be selected, placed in the lamp, and found to give perfect results with a normal $\frac{3}{16}$in. arc. Another candle from the same parcel, in the same lamp, gives only a $\frac{2}{16}$in. arc; whilst yet another finds its place of rest at $\frac{1}{4}$in. Now, to adjust these varying arcs to produce equal light, the top in one case, and the bottom of the lamp in the other case, will have to be weighted so as to adjust the arcs to a normal length. But when lamps are running in series this constant readjustment becomes impracticable; yet without it one lamp would be burning properly, another would be burning, but giving a minimum of light, whilst the other would be flaming—each carbon good in itself, but bad when burnt in series. To what, then, is this irregularity to be ascribed, and how is it to be obviated? The answer is simple. The hydraulic press must be fitted with an atmospheric gauge, capable of measuring at every moment the amount of atmospheric pressure to which the carbon matter is being subjected. The best amount of pressure having been discovered, and care being taken that all the carbons shall be made at that pressure, does it not stand to reason that carbons pressed out at 50 atmospheres would give different results to those pressed at 150? And when we hear, as I have heard, that carbons are irregular, and not so good as they used to be, can it not be well understood how a careless workman in charge of the press, receiving a bucket of pugged material, and passing it through the press without taking any heed of the difference in pressure indicated by the gauge amounting, perhaps, to 20 or 30 atmospheres, vitiates the perfection of the carbon, and does irreparable mischief? The matter lies in the hands of the pressman, who has the gauge for his guide. The man who pugs and prepares the material requires an amount of experience that can only belong to an expert to enable him to grasp a lump of material, and say within 10 atmospheres what the pressure required to pass it through the press would be. Several puggings, and at least three days, are necessary in the preparation of the material. From experience I can say that 120 atmospheres produce good results—nothing under. I have never been able to prepare material requiring a pressure of more than 160; at 120 the carbons leave the press quite warm.

The carbons made by good makers sometimes mushroom. By this is meant the formation of a head resembling a mushroom button upon the extreme point of the negative, which appears to consist of particles of impure matter thrown down from the positive carbon. If this mushroom top increases and does not break off of itself, which it often does, it may touch the centre of the positive cored carbon, in which case the circuit becomes closed and the lamp extinguished. This fault I attribute to impurity of material and the presence of silica, which fastens itself like a glass bead to the point of the negative, and, being a powerful non-conductor, tends to arrest the current. This bead is formed by the electric current searching out the distributed particles of silica in the positive, but without power to vaporise the globule which it deposits on the negative. Pure prepared material is a certain cure.

The lengthening and shortening of the arc, often described as jumping, and which causes the rapid fluctuations in the amount of light, is the characteristic of a bad carbon, badly made with bad material, and owes its existence to a complication of the faults previously mentioned in an exaggerated form.

A word or two upon coppered carbons. Good wine needs no bush, and a good carbon needs no copper. This coating is given to hide a defect; perhaps more than one, since it may be put on to hide a defective surface, whilst a perfect carbon cannot have a bad one. But the chief reason is to get over want of exterior hardness—a very serious fault. Good carbons should be extremely hard on the surface, gradually softening towards the centre, which in the case of cored carbons consists of pure graphite in powder. A carbon insufficiently hardened (hardness being only acquired by repeated saturation and carbonising) burns to a lengthened cone. This is due to the air rushing in to supply the vacuum formed, and which, meeting with little resistance, disintegrates the carbon, and works it to a lengthened cone, presenting only a small point to support the arc, the light in consequence becoming seriously diminished.

PREPARATION OF MATERIAL.

Those who desire fuller information upon the nature of graphite, the various parts of the world in which it is discovered, and other details, may peruse "Spon's" with interest. Two kinds are known, the amorphous and the foliated. The amorphous is quite unsuited to our purpose, and contains often only 45 to 50 per cent. carbon, the rest being argillaceous matter. The foliated under careful test has shown as high a percentage of carbon as 99·679. Crucibles and pencils are made of the amorphous. The foliated graphite which comes to us from Ceylon can be readily obtained in any desired quantity from the Mincing-lane brokers, at prices varying from £16 to £22 per ton, and is well adapted to our purpose. It comes to us in the form of nodules, which require careful selection, and breaking under the hammer in small pieces to extract any quartz discoverable. After careful hand-picking, the graphite must be ground to a fine powder in a mortar or pug mill. After this, the powder should be placed in cast-iron mortars or crucibles, and mixed with chlorate of potash in the proportion of 18lb. of graphite to 1lb. of the chlorate, and 2lb. of sulphuric acid (sp. gr. 1·8) to 1lb. of graphite. This mixture must be moderately heated until the last fumes of the chlorous gas are evolved, and then allowed to cool, when the excess of sulphuric acid should be decanted. Upon this sulphated and oxidised mass pour a small quantity of fluoride of sodium, stir well together, and after the chlorous vapours have passed off, the hydro-fluoric acid set free by the combination of the sodium with the sulphuric acid will convert any silica present into a gaseous fluoride of silica, which passes away in vapour and leaves the mixture pure. This mass should afterwards be thrown into water and well washed, placed again in crucibles, and submitted to a red heat in a furnace. This will cause the whole mass to swell and disintegrate, forming a light flocculent powder floating on the surface, which must be collected and dried. This impalpable powder is yet unsuited for carbon making. A cutting sharp grain must still be imparted to it, and to arrive at this the thin flat plates of gas-retort carbon, taken from the top of the retorts, must be taken and ground fine, in the proportion of one of gas carbon to three of graphite, with which it must be well incorporated in the initial stage, so that its impurities may be equally eliminated. You cannot obtain these thin carbon plates without purchasing the heavier and impurer shoulder-pieces. However, the refuse may be economically employed in furnace heating, as it gives great heat and possesses great durability.

The material so far prepared must be thoroughly dried, and pugged with sufficient caramel or carbonised sugar, mixed with water, to make a moderately stiff paste, placed in ovens or crucibles, and thoroughly carbonised. A material is thus produced possessing all the requisites desirable. It is perfectly pure, and when ground fine possesses a sharp-cutting grain and great hardness, allowing the molecules under pressure to combine to the exclusion of occluded gases, which allows of a perfect homogeneity which could not be otherwise attained. It will be necessary to crush these lumps so prepared in a stone crusher, and after crushing grind them in a mortar mill, passing the powder through a rotary sieve, working from the counter-shaft, having 90 meshes to the lineal inch, or 8,100 to the square superficial inch.

Having described the graphitic material or compound used in carbon candle-making, the manipulation required in the preparation of the stuff, also the way in which the granular condition is acquired, and which experience shows is essential to the production of a good carbon, I now proceed to explain the method of preparation of the saccharine matter employed in combining ˌthe material to form the paste, or pugged stuff possessing the special properties requisite. All materials hitherto experimented upon and used to combine the powdered graphite, such as tar, oils, gums, hydro-carbons, rosin oil, &c., swell the carbons, after being subjected to a drying heat, inducing porosity; and no available after-process is satisfactory. Saturation with carbonised sugar fills the pores, but leads to much waste of time and enhanced expense, both of material and labour. Wellnigh two of the first years which I devoted to this subject were spent in efforts to overcome this difficulty— efforts not made in vain, since not only is the perfect drying of the carbon without swelling now a *fait accompli*; but also the valuable property of contracting under great heat has been realised, so much so, indeed, that allowance has now to be made in the gauge of the die, which, to allow for this contraction, requires to be increased $\frac{2}{18}$th of a millimetre upon a 10-millimetre carbon. Prolongation and intensity of firing increase the contraction. The high conductivity possessed by graphite is, no doubt, due to its peculiar dense condition; but to impart this quality of density we shall have to wait until Nature invites us to work in her own laboratory.

The material I use to incorporate with the prepared graphite powder is the best crystallised lump sugar. Crystallised Demerara might do as well, but the difference in cost is so slight that I prefer to adhere to the former; and, therefore, would cite "Tate's cubes" as the standard. I would urge that no attempts be made to cheapen this part of the process. Molasses and treacle are impure, containing, besides sugar, acetate and sulphate of potassa, chloride of potassium, phosphate of lime and copper, mucilaginous and nitrogenous matters, silica, water, and glucose.

Take, say, 20lb. of Tate's cubes, put them into a boiler large enough to hold four times the quantity, at least; place the boiler on a moderate fire, preferably made of small broken coke, and allow the sugar to melt gradually, without burning, until

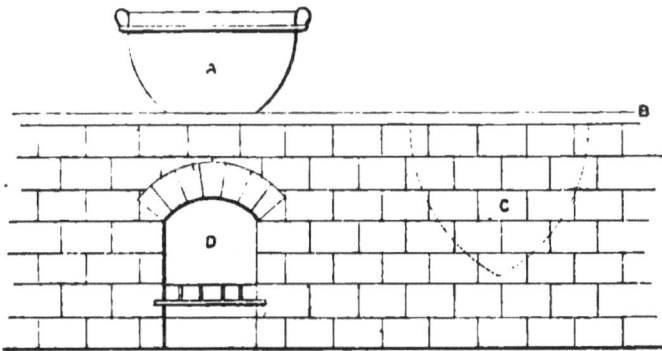

FIG. 1.

it partially carbonises. Careful stirring and watching must be exercised, and it must never be at any moment left, or the sugar is certain to boil over, catch fire, and be wasted. To avoid these *contretemps*, a low range of firebricks may be built (*see* Fig. 1), on the top of which bars, B, of iron are fixed, immediately over the fire, and extending over a sunk pan, C. The boiler A is free to slide upon the bars, so that in the event of a sudden rise it can be pushed over the sunk pan, and any overflow of the contents be recovered without loss.

I have signally failed in the use of beet sugar. When converted into caramel, it decidedly loses the adhesive, sticky property belonging to the cane sugar. Sugar heated to 400° F. loses two equivalents of water, becomes brown, cannot be re-crystallised, and is then known as caramel. At 410° F., the third equivalent of water is set free, and complete carbonisation commences. In this state the spec. grav. is 1·594. Sugar at this stage ceases to rise in the boiler. By observation alone can the point at which the caramel is in the exact state required be determined. As a guide, however, it may be stated that the process is well-nigh completed when the sugar ceases to rise in the boiler ; but make sure that this is not owing to the lowering of the fire. Small volcanic eruptions covering the surface, bursting and emitting vapours which are caused by the third equivalent of water becoming free, is another sign. If any smoke accompanies the vapour, the operation is certainly complete. The process had always better be carried a little too far than not far enough. In the first case, the caramel is only slightly less adhesive, leading to extra pugging; but in the second case the carbons will suffer, and this will not be ascertained until the carbons have been made and placed in the drying-room. To determine the right moment for closing the operation above described needs practice, the appearances shown whilst boiling depending much upon the state of the fire, and being sometimes misleading and puzzling to any but the expert. When the boiler has been duly removed from the fire, add gradually *small* quantities of hot water. Adding cold water in bulk is an experiment not likely to be repeated more than once. Stir well while adding the water, and let this be only sufficient to keep the caramel liquid when cold. When required for use this stock must be diluted with cold water to about the consistency of golden syrup one part, water two parts, possessing a slight stickiness when rubbed between the fingers.

The carbon powder may now be taken out of the storage bin, in quantity sufficient for a pugging, with a pug-mill

6ft. 6in. over all, with rollers 18cwt. each. A 10 horse-power engine is sufficient to drive the mill and the press. About three tin buckets, holding some 20lb. each of the powder, may be taken as a charge, and this weight of material will suffice to form 1,000 10-millimetre carbons. Pug this stuff, adding sufficient of the liquid caramel to form a paste; when pugged, clear out the mill, and transfer the paste to a covered bin, or keep it in the buckets, marking the date when first pugged on the buckets. Continue making these batches, repugging every day until ready for the press. This pug-mill should be kept solely for the purpose of pugging, and not grinding, as the pan could not be so thoroughly cleaned as to avoid the risk of leaving some small fragments of material, which would find their way into the carbon rods, tend to divert the current when used, and give rise to complaints.

A portion of each day may be employed in pugging new stuff, and previous puggings must also have the process repeated daily, each day's batches being kept separate. Three or four days, depending somewhat on the temperature, will bring the stuff to a proper condition for forming into carbons. The paste may be advanced, if required, by placing it in a warm room. By squeezing the paste in the hand the expert can tell the atmospheric pressure requisite to drive it through the press within some 10 atmospheres. This pressure must not be under 120 atmospheres for a 10-millimetre carbon. This same paste, however, that requires 120 atmospheres for 10 millimetres, would in the case of 14 millimetres pass through at 90 to 100 atmospheres. The first few carbons issuing from the press will be no guide, as the pressure will increase until the carbon in the cylinder becomes well compacted; this will occur whenever the cylinder has been emptied for the purpose of changing the die or altering the coring needle.

If the cylinder has been charged, the pressure right, and the carbons issuing from the die, the carbons will require

D

cutting to normal lengths. For this purpose I use a small instrument, which answers the purpose, and cuts off the carbons to a standard length as they issue from the press. It is made of tin, about three-quarters of an inch wide, bent at right angles at D (Fig. 2). The carbon rod A,

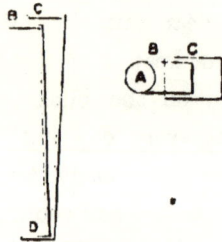

Fig. 2.

shown in section, lies in the grooved top B, the thumb slightly pressing the carbon, while the spring-back cutting part is pressed forward with the fingers ; this closes over the groove B, separating the carbon at the normal length.

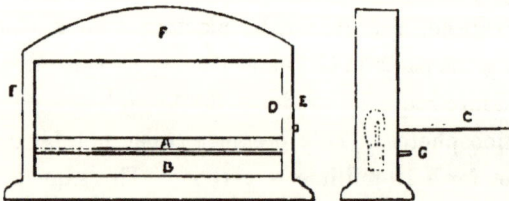

Fig. 3.

As the carbons are thus cut off they must be laid in grooved plates, prepared by a machine (Fig. 3) constructed for the purpose, consisting of a cast-iron frame, with grooved dies ; the lower die B is fixed to the frame EE, whilst the upper die A slides vertically in the frame. This upper die is actuated by a lever handle C, working eccentric chuck-pieces, fastened to each end of the bar, actuated by the lever, and turned by the handle so as by its action to bring the dies in contact. The sheet tin, being laid upon the guide-plate G, is pushed forward through the open dies, and takes the form of the grooves upon lowering the arm; an iron rod at the back also guides the plate in its passage, keeping the grooved plate straight. These dies may be taken out of the frame to give place to others suitable for carbons of other gauge. These grooved plates (Fig. 4) upon which the carbons are carefully laid as soon as they issue from the press, are so constructed and figured that the carbon rods are perfectly protected from the superimposed plates pressing upon them, the plates resting upon each other, the carbons being protected within the grooves; thus a dozen plates filled

SECTION.

FIG. 4.

with carbons may be safely piled one upon another before leaving the pressman. These plates should be made two inches longer than the carbons, of the best charcoal tin, and rather stout. In these corrugated plates the carbons should remain until removed for firing in the kiln, tin being a good conductor of heat, accelerating the process of drying. When the pressman has filled a dozen of these plates they may be placed on a rack in the press-room. Here they may remain until the evening, when they may be transferred to the first drying-room.

I must here describe a fault which may arise by the time this stage is reached. The carbon rods leave the press to all appearance perfect; half-an-hour afterwards every carbon is lined with fine fissures at right angles to the length of the

carbon rod, increasing in size and depth as they dry. [The
fault arises from the die. I had occasion at the time
I first noticed this defect to renew my stock of dies, and

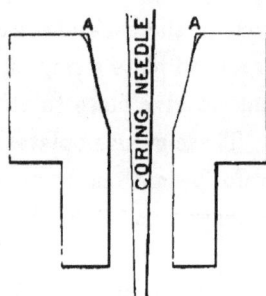

FIG. 5.

gave a model to an experienced workman, who furnished me
with duplicates, apparently in every respect similar to model,
only slightly reduced in gauge. No eye-measurement could

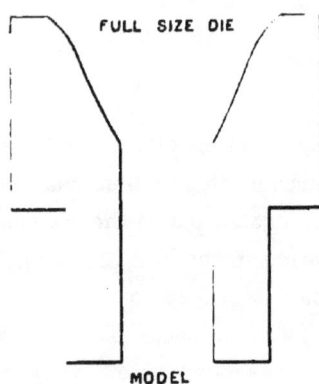

FIG. 6.

detect any difference ; but on slightly filing off the shoulder
at A (Fig. 5) the difficulty disappeared, and the carbons were
perfectly freed from this blemish. A model die for a 10-
millimetre carbon is shown in Fig. 6.

Prior to the discovery of the cause of the lateral fissures referred to I conceived it possible that this fault might have its origin in the speed of the hydraulic pump. This does not, however, appear to affect the carbons, although a speed of about 40 revolutions per minute would seem sufficient.

The second fault we may have to contend with is the sudden appearance of longitudinal fissures, increasing both in depth and width, a few hours after the carbons have been placed in the drying-room. This is due to the fact that the caramel has not been sufficiently carbonised, and still retains some of the water of crystallisation.

The third and last fault which has to be contended with is the curving of the carbon rods as they issue from the press; sometimes only a few carbons are affected this way, sometimes many. It must be caused, I think, by a difference of hardness in the pugged stuff, and, if so, the best way to avoid it will be to have the stuff repugged just before use. The curved carbons must be rejected, whether the curvature is great or small. A curved carbon will aways come out crooked in the firing.

Drying-Rooms.

The most efficient and certainly the cheapest way of heating the drying-rooms is by means of the slow-combustion stove. The temperature can then be maintained, without sensible variation, for quite six hours, if the fires are carefully built up with small broken coke, and without renewal ought to burn for 12 hours before extinction. The first drying-room should be 20ft. long and 8ft. wide, and 8ft. to the springing of the arched roof, with the stove in one corner nearest the door. By this arrangement the furthest end of the room is only slightly heated, whilst increased heat obtains near the stove. Fig. 7 is a ground plan of the first drying-room. A is an iron sliding door, B C D the shelving, framed and built up with angle iron,

as already described. At the end E (Fig. 8), and on the lower
shelf the temperature will remain nearly constant at about
75° F., while the temperature at G will be about 110deg. ; at
H, nearer the stove, the temperature will be 120deg., and at F,

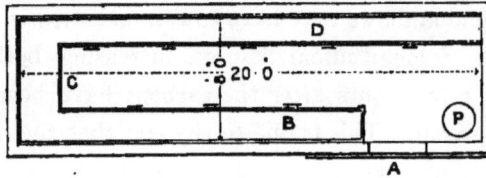

FIG. 7.—First Drying-room (Ground Plan).

140° F. These temperatures will always bear a relative proportion to each other, and never vary much.

The carbons are first placed on the shelves at E, removed to
G, and after 36 hours to F, where they remain for 12 hours;

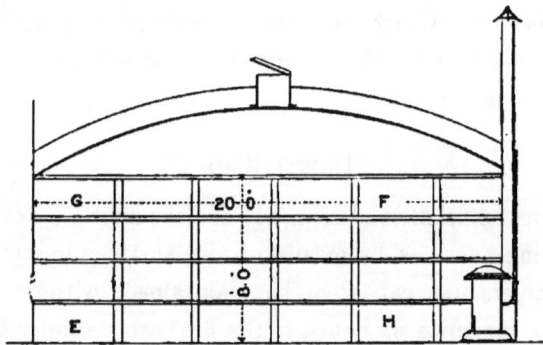

FIG. 8.—First Drying-room (Section and Elevation).

afterwards removed to the second drying-room and kept for
two days at a temperature between 180deg. and 200deg. This
drying-room should be much smaller (about 10ft. by 8ft.),
after which the carbons should be ready for firing. They
ought not to be indiscriminately taken, but should be sepa-

rately handled, as some of the interior carbons may not be sufficiently dry. Their condition is best ascertained by trying to bend them ; if they feel the least inclined to bend they are unfit, and must remain until they are fit. Firing before fitness means breakage.

FIRING THE CARBONS.

I now describe my method of building the kilns. A cleanly but costly system has been tried, but not generally adopted, I believe, of firing carbons by means of gas and atmospheric air combined. The system may be good, but the expense appears to me to be too great. My system is to lay firebrick on the flat, and well grouted with cement; upon which build the two long sides of the kiln B B (Fig. 9), with bricks on edge, well laid in cement, composed of three parts of Stourbridge clay and one part iron grindings. The sludge arising from the grindstone, well mixed, forms a valuable cement, and will not crack under extreme heat. A crack in the kiln admits air, and disintegrates the carbons. The short ends of the kiln are built with bricks on the flat, A A (Fig. 9); the thicker the ends are the better, but 4½in. are sufficient. The kiln must be 2in. longer than the width of two of the plates, and 2in. wider than the length of the plates, which should be 2in. longer than the carbons. This allows for filling in with sieved ashes, which should be kept perfectly dry ; any trace of moisture will show itself in the partial disintegration of the carbons.

The lines at C C (Fig. 9) show the direction in which the carbons are to be packed. The ends A A are made thick to avoid any lateral heat ; the fires being made at D D tend to keep the carbon straight. Any heat in the direction of A A would inevitably twist them. After the kilns are packed and covered over to a depth of 2in. with the ashes, the surface had better be covered with a sheet of tin, and built in with firebricks, well filled in between the joints with the prepared cement, and pointed before using the kilns. After constructing the kiln a fire should be made in and around, so that it may be perfectly dry when used. The trays in which the carbons are fired must be kept solely for that purpose. The furnace con-

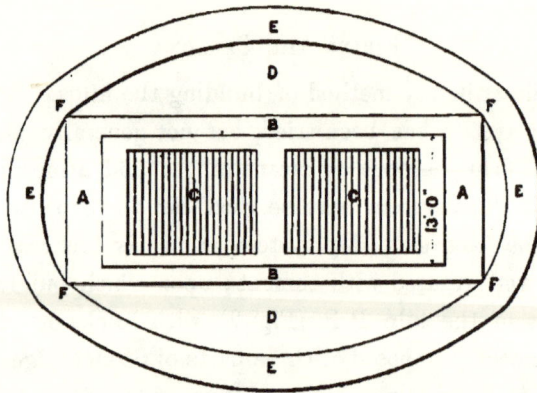

FIG. 9.—Kiln (Ground Plan).

sists of a ring of firebricks, built as shown at E E E E (Fig. 9), with interstices between them of about 2in., E E. (Fig. 10). This circular wall should impinge upon the kiln at F F (Fig. 9), to keep the heat from the sides A A. The bricks are laid loose

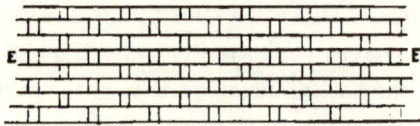

FIG. 10.—Brick Wall of Furnace.

without cement, and are built to the height of the kiln. The fires must be lighted at D D, gradually adding coke as the fire advances, eventually increasing the height of the walls about a foot higher than the kiln. If the fire was lighted the first thing in the morning, it should not be piled over the top until the evening, and this done only by covering with coke and banking over with cinders and clay. Allow this fire to die out, when the carbons will be found, on opening, when cold, sufficiently carbonised for the after-process.

The second or final firing has for its object the contraction of the carbon molecules, and increase in density and conductivity. The first firing, however, must be sufficiently prolonged to effect the carbonisation of the caramel, otherwise when placed in the hot saturating bath the carbons would become partially disintegrated. The second firing should be conducted as follows : Bank up as before mentioned in the evening; on the following morning rake out the lower holes all round, and keep a strong fire all day all over the kiln. Bank up partially in the evening, and more completely the following morning. Three days after, on opening, the kiln will be found at a white heat. When it is desired to withdraw the carbons, pull down the retaining wall, and clear away the fire and accumulated ashes, and expose the kiln to the air. Do not open before you can bear your hand on the outside. The top bricks may be taken off, and the kiln emptied, if the carbons are not too hot to hold. I should mention that straight carbons will twist if exposed to the air hot from the kiln ; but if once allowed to cool, no after-firing has the least effect upon them. This form of kiln gives out an intensity of heat sufficient to melt wrought-iron, is very easily regulated and modified by blocking up more or less the interstices, and the firing can be prolonged for any length of time by banking up with brick clay, which in itself partakes of the nature of a slow-burning fuel.

The condition of the carbons as they leave the kiln may, in some measure, be determined by the sound emitted when struck one upon another. This sound should be distinctly metallic, the more so the better. This metallic ring increases in the finished carbon to the clearness of a small silver bell; it

E

is only heard in good carbons. A carbon, however, is not neces-
sarily perfect, although the ring may be perfect. It is very
desirable to know in what condition the carbon leaves the
kiln, to be able to determine what it lacks, and what it requires
to perfect it. The feebleness of its metallic ring will indicate
that it is soft. If you insert the end of one of these carbons
in a clear hot fire until it is red hot, and on taking it out
observe it closely, you will find it smoulder for quite five
minutes before it dies out. If you scrape this end with a knife,
much carbon powder will detach itself. Observe also the
colour of the powder ; if reddish, the carbon is impure. The
colour of the burnt part of a good carbon is a bluish steel-grey.
This holds good for all the best foreign carbons. If the carbon
just taken from the kiln had been put in circuit, the current
would have made it red hot from end to end, but no arc would
have formed for want of conductivity. By imparting increased
density to the carbons, these symptoms of an unfinished carbon
are altered, and give place to easily recognised signs, which
are always the characteristics of a good finished carbon, viz.,
rapid extinction of the heat after firing the end, and hardness of surface, which becomes as hard as a rod of iron, and
does not allow of the surface being disturbed or injured by
scraping.

SATURATING.

Now, to effect this change, the carbons must be saturated,
and the operation is carried out as follows :—An iron tank, A
(Fig. 11), is provided, partitioned off at one end to contain hot
water, and the tank is placed upon a range, with two fireplaces, B B. Fires being lighted, the tank must be filled to
the depth of 12in. with water and a sufficient quantity of sugar
to form a syrup. The boxes, D, constructed of tin or zinc,
with perforated bottoms, are filled with the carbons packed
vertically, and placed in the tank. After boiling half an hour

the boxes are removed, and placed upon iron bars, E E, to drain. When cold replace them in the tank, and repeat the process three or four times. Drain them well, and wash the carbons freely in the hot-water end. Remove them when cold (by which the liquor is drawn into the centre) into the third drying-room, and keep them there for 24 hours at a temperature of between 180° and 200° F., after which the whole process may be repeated until the carbons show the characteristics described above of a finished carbon.

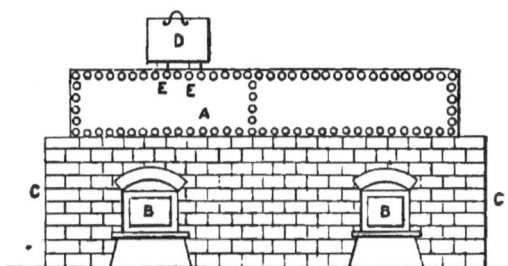

Fig. 11.—Saturating Tank.

Some time ago, in hunting over patents at the Patent Office, I came across a patent (provisional) for saturating carbons. I have never tried it, but I give it here for what it is worth, and I hope not to the injury of the patentee. Steep the carbons in a dilute solution of calcic chloride, afterwards into a saturated concentrated bath of potassa chlorate (at a boiling temperature), dry afterwards by heat, and whilst hot steep in another bath of fused sodic nitrate, and wipe clean whilst hot. What effect this Turkish bath may have upon the carbons I am not prepared to say, but I think it might be worth a trial.

CORING.

Numerous experiments on the different modes of coring enable me to say that the method of coring is of less importance than the condition of the material. A hard core is useless. The best possible form is that of a compressed powder, breaking up easily with the point of a knife ; but if too softly packed the current is apt to throw it out, and leave a vacant place which degenerates into a hole, and the light becomes diminished.

Every manufacturer of carbons has, doubtless, a way of his own of coring carbons. I have experimented on three very different methods, and I now explain them. The most certain, the best, but I must also add, the slowest way, is as follows :— Take the finished carbon, and fit a small funnel with a nozzle large enough to fit close exteriorly, partially filling the funnel with graphite powder, purified, and fired at a white heat in a graphite crucible, and then tamp the powder. With a little practice this is quickly done, but scarcely quick enough for commercial purposes. The core, however, by this method can be made hard or soft, as desired. The second method is performed much more rapidly by using a specially-constructed coring machine. Before, however, describing it more particularly, I must mention certain difficulties. If the powdered graphite is mixed with water, on applying pressure with the screw the water separates, rising above the piston, whilst the powder hardens with the pressure and becomes like stone. Powdered graphite mixed with caramel fills the carbons easily, but after firing the carbon becomes harder than the ordinary lead pencil, and is spoilt. If oils are used, or liquid gums, the core is partially driven out in drying, the medium having a tendency to swell, so that the carbon is but partially filled, and the paste becomes attenuated and honeycombed when it should be close and solid, and the carbon may flicker. The only medium that answers, as far as my experience goes, is heavy mineral oil, with a specific gravity of 1·82. Mixed with this, I have not found the carbons swell nor get too hard when fired. The third and last method which I wish to mention is the gravitation method. A cask partially filled with the powder is placed under a tap of running water. This cask communicates by means of pipes with boxes packed with carbons, and as the cask runs over it fills the boxes, and the fine powder floating in the water gradually subsides, filling the core-holes. The cask requires stirring up occasionally. Where many thousands have to be cored this process will be employed, the system being automatic. The surface water should be decanted when the carbons are completely covered, and the boxes removed to the

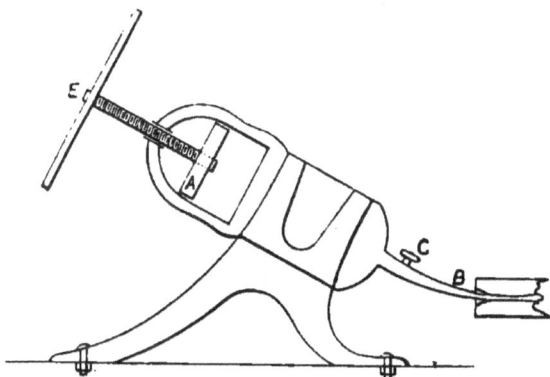

FIG. 12.—Coring Machine.

third drying-room, and left to dry, when the carbons may be withdrawn, wiped clean, and dried upon the shelves.

The coring machine (Fig. 12) explains itself. A is the piston actuated by a screw revolving upon the wheel E. C is a tap fixed to the nozzle B, which, tapering, enters the core-hole of the carbon. The finger is placed at the end of the carbon. Two or three turns of the screw will put sufficient pressure upon the material to fill several carbons, by simply turning the tap at C, which must be turned on and off as required. This method of coring must be adopted after the carbons are saturated and have been washed for the last time, the core hole being wet and allowing of the free passage of the paste. The pointing of the carbons is most easily effected on an emery wheel, revolving, as all emery wheels should to be of any use, at a high speed, 1,200 to 1,500 revolutions per minute.

I now come to an important matter, the construction and adjustment of the dies and coring needle. The pressure to which this needle is subjected when embedded in the pugged stuff is enormous ; 120 atmospheres have to be resisted. On more than one occasion I have had the wings of my coring needle crumple up like paper ; and to prevent this a proper form must be employed. A A (Fig. 13) represent the sides and base of the cylinder, E E a loosely-fitting die resting upon this base. C C is the die proper.

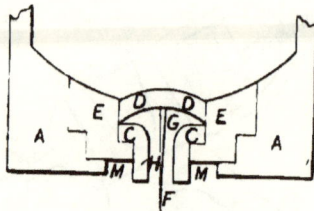

Fig. 13.—Die and Needle.

The coring needle F must be screwed into the wings D D, which are four in number. The size of the core-hole will be determined by the size of the needle at H, no matter what the thickness of the shoulder. The die proper C C is sunk below the outer die E E, so as to allow the wings D D to find a bearing against the outer die E E. The wings are made sharp-edged above, gradually thickening to 4 mm. at the base.. The coring needle, when properly adjusted, will always be central, unless the wings get twisted by the pressure. The motive for making the die E E fit loosely is to facilitate the removal of the needle and the inner die. When necessary this is most effectually done by taking a copper tube, placing it at M M, and striking it with a heavy hammer ; this will bring up

the pugged stuff and the dies, instead of picking out the
pugged stuff; the proper die will also be preserved from
damage.

The hydraulic press (Fig. 14) which I had constructed was
made of steel, but this is not necessary, as cast iron will do
perfectly, since it will not be required to stand a greater pres-

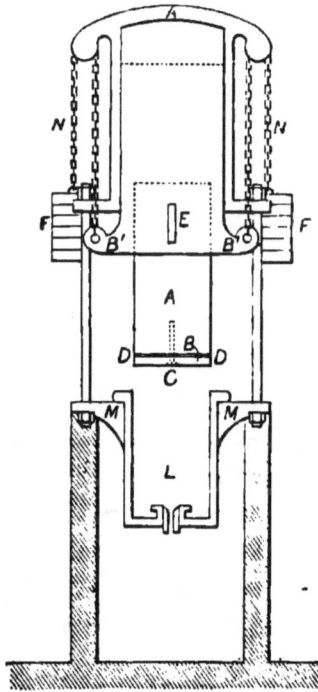

Fig. 14.—Hydraulic Press.

sure than 200 atmospheres; the ram should be 12in. diameter,
and the plunger quite 8in. diameter, if it is required to turn out
20,000 10mm. carbons per week. With a cylinder holding the
pugged material 10in. deep, the plunger is keyed into the ram
at E, so as to be removable, if necessary. The weights F F
consist of iron slabs, 1cwt. each, suspended from the chains

N N. Five slabs on either side are quite sufficient to raise the piston. The chains are fastened to the lugs B′ B′, which are part of the casting. The cylinder L with collar C drops into the framing M M, fitting tight, to prevent rocking. These parts must be made with greatest nicety. When the piston A has descended to its greatest depth it is sure to jam, unless the cylinder is strongly fixed. This may be done in many ways, and need not be explained, but must be attended to. I have employed a leverage of at least 50 tons to separate the piston from the cylinder. Not only must the cylinder be immovable, but the piston must be treated in the following manner. The piston must be lengthened, say $1\frac{1}{4}$in., by a piece screwed on to the piston end by the screw C, the head well let into a counter-sink. The addition must be a shade larger in diameter than the piston proper, with a small groove, D D, at junction. Before the piston enters the cylinder a little heavy lubricating mineral oil should be rubbed round the piston. After I had adopted this very simple plan the pugged stuff never clogged the piston higher than the groove D D, and the piston always rose unaided.

The manner of setting up the press needs no description ; but being some five tons in weight, requires a good foundation. A word about the pumps. Sometimes the high-pressure pump gets out of order, a fact known by the arrested motion of the descending carbon-rod. Either the suction valve, or the cut-off valve, is out of order, or leaks. To dismount the pump is needless. Empty the tank, and wash it out with hot (not boiling) water. Throw washing soda, or soft soap, into the water, making it pass through the pump to cleanse the valves, which in time invariably become clogged with grease. The ram has constantly to be greased to keep the collar soft and pliable. Sometimes the piston of the pump requires repacking. Besides these annoyances—easily remedied—the pumps should work indefinitely.

DESCRIPTION OF PLANT REQUIRED.

Having fully described the various methods employed in the manufacture of carbons, I am led to the belief that the description of the plant required will be helpful, also a ground plan

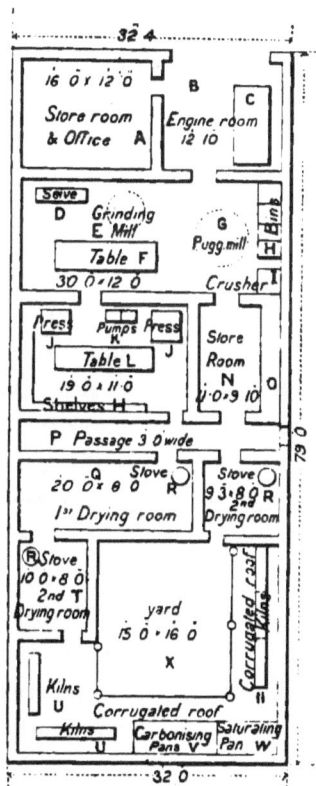

FIG. 15.—Carbon Manufactory.

of premises (Fig. 15) suitable to an output of 20,000 carbons per week, showing the various rooms required and their general arrangement. In the ground plan the rooms, as will be observed, are large enough to contain two presses and two pug-mills, so that a larger number than that stated could be manufactured without increased accommodation. The yard, however, would not be large enough for over 20,000. Of course, it will be seen that the plan is drawn rather for the purpose of showing the size of the rooms required and the most convenient method of arrangement, which may be modified as circumstances require. I place my engine-room facing the open for the convenience of installing both the engine and the general machinery. The pug-mill is best situated near the engine-room, whence the motive power can be directly supplied without the intervention of the counter-shafting, by means of which all the lighter machinery will be driven. The store room N receives all the material required for the processes. From this store-room the material passes at once to the crusher I, from there into the grinding mill E, and thence it is removed into the yard to undergo the purifying process, and to the carbonising pans V. After being pugged and recarbonised it is ready for crushing, grinding, and pugging, and it is placed in the bins H until required. These bins are situated for convenience close to the pugging mill. When ready the material is taken to the table L, close to the press J, whence it issues in the form of carbon rods, which are placed in their separate trays and removed to the shelves H. From here the carbons are removed to the first drying-room Q. After drying the carbons in the first drying-room they are to be removed into the second, T ; and it will be observed that by the adopted arrangement every process, beginning from the engine-room, passes gradually through the rooms. From the second drying-room they are taken to any of the vacant kilns U U to be submitted to the first firing. After this they are at once removed to the saturating pan or tank, W, whence they are again removed to the third drying-room. After the saturating process is complete and the carbons are dried, pointed, and cored, they are removed to the store-room A. The writer has erected premises upon this model, and his study has been to avoid unnecessary carriage of material from place to place, especially in the case of the unfinished carbons, which in the early stages are tender and easily broken.

Having said so much upon the subject, it will be desirable to state the various speeds which I have proved most suitable for carrying on the work. Given the speeds, it will be easy to apportion the diameters of the pulleys, with reference to the diameter of the main driving-wheel, to arrive at any speed required.

It will be well to set down at first a 16 horse-power engine capable of driving two pug-mills and two presses ; but this will, of course, depend upon the scale upon which it is proposed to establish the works. The pug-mills may be allowed to make twelve revolutions per minute, not more. Even at that rate of speed the material in course of pugging clinging to the rollers will occasionally be ejected from the pan. To avoid this, special provision must be made, such as a shield of stout sheet-iron, fastened with bolts and nuts to the periphery of the pan, and rising 18 inches. By the adoption of such a shield the material falls upon it and returns into the pan. A sliding door should be fitted to allow of access to withdraw the pugged material. It will be also necessary to provide a collar either of tin or (preferably) leather to surround the central axis of the mill, and prevent the admission of the graphite powder, which otherwise grinds away the axis, and makes it heavier to work. The scrapers must also be specially designed, and require to be made strong enough to resist the strain, which is enormous. The pug-mill should be fixed upon a very solid foundation, built of concrete, with stone pillars, to support the iron legs of the mill, which must be bolted down to the stone-work and run with lead. The grinding mill must be similarly erected, running at a speed of 12 revolutions per minute ; but to avoid the waste of powder and the dust arising from the process it will be imperative to cover in the pan with an iron (sheet) cover, fitting round the top, clearing the rollers, and suspended by a cord and pulley, to be elevated, when the ground powder requires to be taken out for sieving. The rotatory sieve should work off the counter-shaft with a speed of about 60 revolutions per minute. The meshes of the sieve should be of brass, and must be 90 to the lateral inch, or 8,100 meshes to the superficial square inch. It will be inadvisable to pass the ground powder through the machine direct; it should be passed first through a hand sieve, with 40 meshes to the lineal inch, to extract any small fragments which may have escaped being ground. Unless this is done, the feeding rollers of the sieving machine will be arrested, and require constant attention. The delicate wire sieving will also be torn, allowing the passage of these coarser particles to mingle with the finer powder. In order to be certain that this sieve is sound, it will be necessary at all times, before clearing the middle box, to carefully pass some of the powder between the finger and thumb so as to detect the presence of grosser particles. As the machine is closely covered in, and cannot be examined without some trouble, this will be seen to be clearly necessary; and if grit is met with the machine must be dismounted and the sieve renewed or otherwise repaired. The hydraulic press and pumps have been described. The crushing machine, so long as it effectually does its work, may be of any type, but a disintegrator is not necessary nor advisable.

It may be as well to make some mention of the arrangements of the yard, a portion of which must be covered in ; and this had best be done with sheets of corrugated iron, 7ft. in length. The plan shows the proposed method of arrangement sloping inwardly from the outer walls. By means of this roofing the kilns will be protected from the rain, whilst the centre of the yard is open, affording a suitable place for treating the crude material during the process of purification with the acid, and allowing the escape of noxious fumes and generated chlorine gas. Storage room also is required for coal and coke and clay.

The carbonising pans and range arrangements (Fig. 16) consist of a series of cast-iron vessels, A A A, with close-fitting lids, built into a range with suitable fireplaces, C C C. In these vessels the purified material can be readily carbonised, the first process of purification being carried out in the open yard, as well as

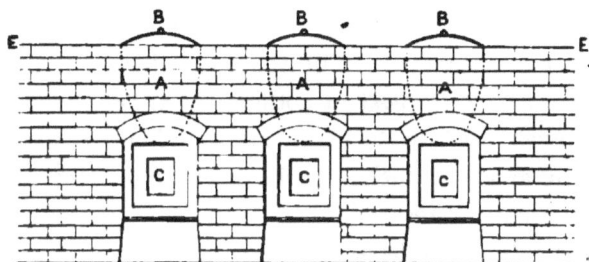

Fig. 16. — Carbonising Pans.

the washing. It would be convenient and inexpensive to illuminate the premises with incandescent lamps instead of gas, with the dynamo placed in the office to avoid the carbon dust.

Cost of Plant.

I now propose to deal with the cost of the plant required for an output of 20,000 10 mm. carbons per week, leaving it to those immediately interested to make their calculations for any larger number.

The practicability or otherwise of establishing factories in this country for the manufacture of carbon candles must depend upon the relative values of material and labour here and abroad. A close investigation of the probable cost seems absolutely necessary, nor would the subject be complete without some knowledge of the kind. It would be unfair to take extreme prices either way. The maximum would obtain in establishing works near London, where price of fuel alone would

militate against success, and the rent of premises even more so.
I will not, therefore, base my calculations upon extreme prices,
but place the works somewhere in the vicinity of gas-works,
where coal and coke are cheap. If my prices are too low,
perhaps the margin of profit may, even with the addition, prove
satisfactory. It appears to me a matter of small question
where the works are established—the cost of transit of the
material employed need scarcely enter into calculation ; but the
price of coal and coke is an important matter, overriding
every other consideration. In the neighbourhood of coal-pits
breeze can be had for next to nothing. This, mixed with clay,
a fuel much used in Belgium and the frontiers of Prussia,
makes an excellent fuel suitable to every purpose save the
engine, which requires coal.

The calculations I have made are based upon an output of
20,000 carbons cored per week, 10 mm. diameter and 10in.
long, or 15,000 as above and 15,000 negative solid carbons
6in. long and 6 mm. diameter.

The cost of the machinery delivered from the works will be
as follows :—

	£	s.	d.	£	s.	d.
1 hydraulic press	100	0	0			
1 pug-mill	60	0	0			
1 grinding mill	35	0	0			
1 sieving machine	25	0	0			
1 crushing machine	12	10	0			
3 coring machines	25	0	0			
1 corrugating machine	8	10	0			
16 H.-P. engine	180	0	0			
				446	0	0
Iron-work, tank, cast-iron crucibles, furnace tank	60	0	0			
Counter-shafting, belting, &c.	50	0	0			
Adapting premises, &c.	120	0	0			
Kilns, furnaces, iron roofs	45	0	0			
				275	0	0
Rent, rates, and taxes				50	0	0
				£771	0	0

Cost of Materials.

These calculations are based upon an output of 20,000 positive cored carbons per week, as before mentioned, upon which a week's labour has been charged.

	£	s.	d.	£	s.	d.
900 lbs. graphite, at £18 per ton	8	2	0			
300 lbs. gas carbon, at 25s. per ton	0	3	6			
2,400 lbs. sulphuric acid, at 1d. per lb. (half of this is saved)	5	0	0			
60 lbs. chlorate potash, at 7d. per lb ...	1	15	0			
240 lbs. Tate's cubes, at 3d.	3	0	0			
3 tons coal (engine), at 12s.	1	16	0			
8 tons coke (firing), at 8s.	3	4	0			
				23	0	6

Cost of Labour.

	£	s.	d.	£	s.	d.
1 superintendent, per week	5	0	0			
1 engine-driver, at 25s.	1	5	0			
1 yard man, at 20s.	1	0	0			
1 man (pugg), at 18s.	0	18	0			
1 boy „ at 12s.	0	12	0			
1 boy (yard), at 12s.	0	12	0			
1 boy (general), at 10s.	0	10	0			
3 girls for coring, at 8s	1	4	0			
				11	1	0
				£34	1	6

The price of foreign carbons is difficult to arrive at. Large orders naturally command large discounts, hence the published prices are no guide to the consumer. In some cases I have known 50 per cent. discount. Even more, I have no doubt, has been allowed. The reader must therefore find out for himself what the best foreign carbons can be supplied at.

Here I take the liberty to fix the wholesale price, net, at £5. 10s. per thousand cored 10 mm. carbons, as the market price, below which I doubt much if they will be ever sold, the competition being very small, and the demand daily increasing.

I have not taken notice in these calculations of interest on capital or deterioration of plant.

ESTIMATED PROFIT.

	£	s.	d.
Cost of 20,000 foreign cored carbons, 10 mm. diameter, 10 inches long, at £5. 10s. per thousand	110	0	0
Cost of material for 20,000 home-made carbons £23 0 6			
,, labour ,, ,, ,, 11 1 0			
	34	1	6
Profit on home-made carbons	£75	18	6

Thus showing a gross weekly profit of no less than £75 18s. 6d. on 20,000 carbons, the profit increasing largely with the diameter of the carbon.

PRINTED BY GEORGE TUCKER, AT 1, 2, & 3, SALISBURY COURT, FLEET STREET, E.C.